Beginner's Guide to...
Correlation Analysis

Learn The One Reason Your Correlation Results Are Probably Wrong

(and what you need to do to fix them)

Copyright

Beginner's Guide to Correlation Analysis (2nd Edition)

By Lee Baker

Copyright 2018 Lee Baker

Amazon Paperback Edition

Thank you for purchasing the **Beginner's Guide to Correlation Analysis** (2nd Edition).

You are welcome to share it with your friends.

This book may be reproduced, copied and distributed for non-commercial purposes, provided the book remains in its complete original form.

If you enjoyed this book, please return to your favourite eBook retailer to discover other works by this author.

Thank you for your support.

Contents

Introduction

What are Associations and Correlations?

Statistical Correlations

Statistical Associations – Measurements Within Categories

Statistical Associations – Counts Within Categories

Your Correlation Results Are Probably Wrong!

Bonus: Automating Associations and Correlations

Summary

About the Author

Claim Your FREE eBook Now!

Leave a Review

Claim Your FREE eBook Now!

Correlation Is Not Causation is the sister book to **Beginner's Guide to Correlation Analysis**, and shows you how to avoid the 5 correlation-causation traps that even the pros fall into.

Download your **FREE** copy right here:

http://bit.ly/2FetzGT

Introduction

Most books about correlations are difficult to read. In fact, most books about statistics are difficult to read and it leads you with the distinct impression that statistics is hard. I'm not going to tell you that it isn't, but I will tell you that **statistics is probably a lot easier than you think it is**.

Over the years, I've taught statistics to scientists of all sorts of backgrounds. Nurses and surgeons, technicians and pathologists – even marketers, business executives and sales people. They all had two things in common:

1. they insisted that they didn't understand statistics, and
2. they all had the light-bulb moment when they suddenly realised that getting the result they were looking for wasn't so hard after all.

You see, statistics isn't about statistical tests. Really, it isn't. It's also not about 'pressing buttons until I get a p-value that looks about right', as one renowned, world-class researcher once said to me.

It's about understanding data, what you can and can't do with it, and how to extract the right kind of information from it. Statistics is just a toolbox that you can use to reach the story locked inside the data.

So here in this book, we're going to look at correlations and associations with a focus on understanding the data, how to work with it, choose the right ways to analyse it, select the right tools from our statistical toolbox and how to interpret the results in a way that is easy to understand.

We're going to start out by learning the difference between correlations, associations and statistical relationships. From there we're going to delve a little deeper into correlations and two different types of associations, learning how to arrange your data, how to plot it and how to interpret the plot. We'll learn how to analyse the data to get a better understanding of what it's trying to tell us, which statistical tests to use in various situations, and we'll learn to recognise which statistical results are the ones that we need to pay most attention to.

Then I'm going to throw a spanner into the works by telling you that you can't trust any of it!

We'll then visit the world of multivariate statistics, where we'll start to understand why we need these more powerful statistical tests to get the correct answers out of your data.

Finally, there's a bonus chapter, where I'm going to show you how you can visualise all the relationships in your data in minutes rather than months (no cheating and taking a sneak peak – you're just going to have to wait…).

Learn More

There is a **Resources page** that accompanies this book, where you will find reviews and links to other **books**, **blogs**, **video courses** and all sorts of other useful stuff.

This page is updated regularly so you'll always know where to find the best the web has to offer.

Visit the resources page here:

- http://chi2innovations.com/blog/discover-stats-blog-series/one-reason-your-correlation-results-are-probably-wrong/

What are Associations and Correlations?

Did you know that statistics can *never* prove that there is (or is not) a relationship between a pair of variables?

If that's the case, then what is the point of statistics, I hear you ask...

Well, **statistics is the study of uncertainty**. If you've already proven beyond all doubt that a relationship exists between *this* and *that*, then there is nothing to be gained from a statistical analysis. It is only when there is *uncertainty* about the relationship that we can learn something by using stats. It is for this reason that statistics can neither prove nor disprove the existence of a relationship. It can only tell you how likely or unlikely that a relationship exists.

So what is a statistical relationship?

When you can phrase your **hypothesis** (question or hunch) like the following, then you are talking about the relationship family of statistical analyses:

- Is smoking *related to* lung cancer?
- Is there an *association* between diabetes and heart disease?
- Are height and weight *correlated*?

Typically, the terms **correlation**, **association** and **relationship** are used interchangeably by researchers to mean the same thing. That's absolutely fine, but when you talk to a statistician you need to listen carefully – when she says correlation, she is most probably talking about a statistical correlation test, such as a Pearson correlation.

There are distinct stats tests for correlations and for associations, but ultimately they are all testing for the likelihood of relationships in your data.

When you are looking for a relationship between two continuous variables, such as height and weight, then the test you use is called a correlation test. If one or both of the variables are categorical, such as smoking status (never, rarely, sometimes, often or very often) or lung cancer status (yes or no), then the test is called an association test.

Learn More

Just a little reminder about the **Resources page** that accompanies this book.

If you want to learn more about correlations and associations (and get even more **FREE stuff**), this is where you'll find it:

- http://chi2innovations.com/blog/discover-stats-blog-series/one-reason-your-correlation-results-are-probably-wrong/

Statistical Correlations

If there is a correlation between one variable and another, what that means is that if one of your variables changes, the other is likely to change too.

For example, say you had a **hypothesis** (a theory or hunch) that there was a relationship between age and percentage of body fat; that as you aged your percentage of body fat increases. We could go out and collect some data, like this:

Age (Years)	Body Fat (%)
23	9.5
23	27.9
27	7.8
27	17.8
...	...

You would do a **scatter plot** of age against body fat percentage, and see if the **line of best fit** (aka a **trend line** or a **regression line**) is horizontal, vertical or if it has a trend. If the line is horizontal or vertical, that means that as one variable changes the other does not. In other words, there is not a relationship between them.

On the other hand, if the trend line is not horizontal or vertical, then we can say there is a trend, like this:

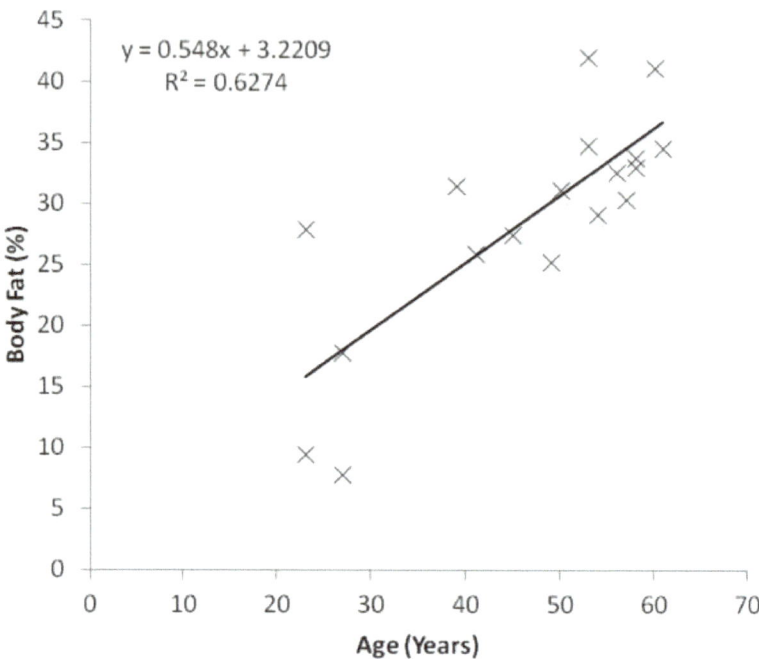

Scatter Plot Illustrating a Positive Correlation between Age and Percentage of Body Fat

In this example, the most appropriate trend line is one that is linear (but in other data that is not necessarily the case), and there is a positive correlation between age and percentage of body fat. That is, as you get older you are likely to gain increasing amounts of body fat. Ah, the joys of getting old…

Notice, though, that there is no suggestion that getting older *causes* you to gain extra body fat. You should probably know by now that correlation does not necessarily imply causation. To find out exactly why this is the case, don't forget to download your FREE copy of **Correlation Is Not Causation**. I'll remind you at the end of this chapter where you can get a copy.

On the chart, we have the **equation of the line**, which also includes the **gradient** (in this case, 0.548). The gradient tells you by how much the y-axis changes for each unit change in the x-axis. Here, we can say that body fat increases by a little over 0.5% every year.

A common misconception that the inexperienced make about correlations is that they assume that as long as the trend line is not horizontal or vertical, this means that there must be a correlation between the variables. This may or may not be true. You see, it depends on how far the plot points fall from the line of best fit. If they are all very close to the line, then you can be quietly confident that there may be a correlation, while on the other hand if the plot points are scattered then you should be less confident about a correlation.

You can assess your level of confidence by calculating the **R^2 value** of the data. All statistics programs will be able to calculate this for you, even Microsoft Excel.

The R^2 value (given on the chart above) is a measure of how far the plot points fall from the line. Briefly, the vertical distance from each plot point to the trend line is calculated. If these distances are small and consistent, the R^2 value will be large (close to 1), whereas the larger and more variable they are, the smaller the R^2 value will be (close to 0). R^2 values are also often displayed as percentages.

In the graph above, the distances of the plot points from the regression line are small and mostly consistent, but with a few points that deviate appreciably. The R^2 value therefore tells us that there is a strong positive relationship between age and body fat (an R^2 value larger than 0.5 can be crudely categorised as 'strong').

We can't tell though just from looking at the R^2 value whether the relationship between age and body fat is statistically significant. For that, we need to run a statistical correlation test.

If the relationship between the variables is linear and a change in one variable is clearly associated with a proportional change in the other, then a **Pearson**

Correlation test is appropriate. If the relationship between your data is such that a change in one variable does not necessarily indicate a proportional change in the other (i.e. they do not change together at a constant rate), then **Spearman's Correlation** would be more appropriate.

For our data, the relationship is such that change in age corresponds to a proportional change in body fat (the best-fit line through the data points is a straight line), so we run a Pearson Correlation, like this:

Regression Analysis: Body Fat versus Age

The regression equation is
Body Fat = 3.221 + 0.548 Age

Predictor	Coef	SE Coef	T	P
Constant	3.221	5.076	0.63	0.535
Age	0.548	0.106	5.19	6.16e-05

S = 5.754   R-Sq = 62.7%

This is a typical output that you'll find from many statistics programmes, although some may also output pages and pages of other stuff too. Some of it might be useful, but most of it won't be, and to get started you need to look for the results that look like the table above.

OK, first thing to notice is that you get the equation of the line again. Check that it's the same as the equation of the best-fit line that we generated in Excel (spoiler alert: it is).

Next there is a table that contains a lot of numbers, some of which we don't need (they are used to calculate other numbers in the table, and seasoned statisticians start to feel uncomfortable if these other values are missing, so we humour them and leave them in...). The first important thing about the table is to note the coefficients. We use these to form the equation of the line. The coefficient of the Constant is the intercept and the coefficient of the variable Age is the gradient.

I'm going to skip right over the **Standard Error of the Coefficient** (a measure of the variability of the plot points) and the **T statistic**, and jump right to the p-value. The **p-value** of Age (6.16e-05) is very small and crucially is less than 0.05, and this tells us that the correlation between Age and Body Fat is **statistically significant**.

Below the table are a couple of **goodness-of-fit** test results. We have already seen the R^2 value (again, check that it matches the R^2 that we generated in Excel – another spoiler alert: it is), but the S value – the **Standard Error of the Regression** – is much more informative. The Standard Error gives us the average distance (arithmetic mean) that the plot points fall from the regression line, and is in the same units as the y-axis. Here, S tells us that the average distance between the plot points of body fat and the regression line is 5.754% body fat, and we can use this to show the average variation of the plot points on the graph, like this:

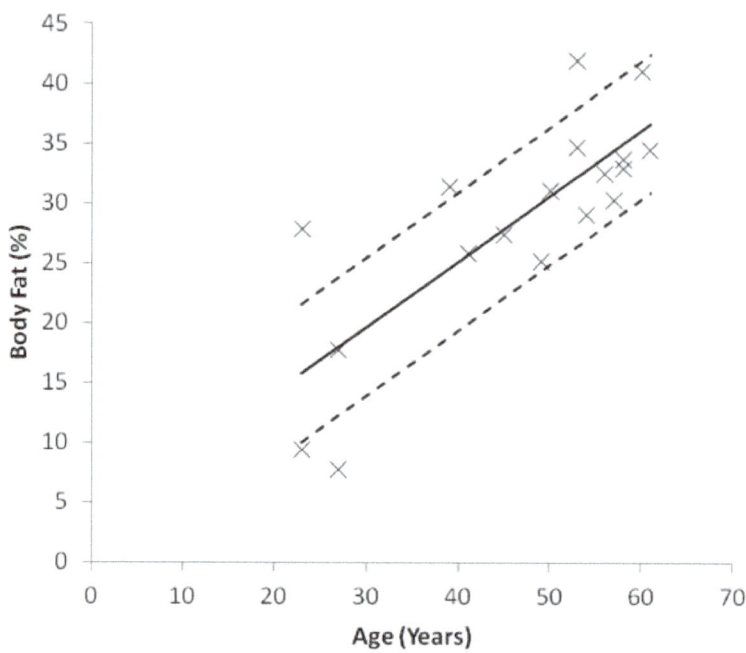

Scatter Plot with Line of Best Fit and Standard Error of the Regression

So, we started out with some data about Age and percentage of Body Fat, and we wondered if there might be a relationship between them. By plotting them against each other on a simple scatter plot in Excel, we could add a best-fit trend line and have Excel calculate the gradient, the equation of the line and an R^2 value that tells us how far the plot points fall from the line – an indication of whether we should be confident about a correlation. We discovered that by using a correlation test we could find out if the correlation is statistically significant, and that the Standard Error of the Regression gives us more information about the spread of the plot points.

And we did all this with just a few clicks of the mouse...

Next, we're going to look at statistical associations and see if we can pull off the same trick!

Test Yourself

- Looking at the best-fit line in the scatter plot above, what would be your expected increase in body fat percentage over the next decade?
    - [Clue: rather than guessing from the regression line you could calculate it from the gradient in the equation of the line].

Learn More

Still not visited the **Resources page** yet?

Well how about a little tempter? Visit the page to claim the book **Correlation Is Not Causation** absolutely **FREE**!

And if that's not enough, there's even more FREE stuff there to claim. Hop right over. You'll find it here:

- http://chi2innovations.com/blog/discover-stats-blog-series/one-reason-your-correlation-results-are-probably-wrong/

Statistical Associations – Measurements Within Categories

As we've just seen, statistical correlations look at relationships when both variables are continuous. When one of the variables is labelled in categories, then we are dealing with statistical associations, specifically the measurement of a variable within categories.

Let's have a look at an example from Charles Darwin. In 1876, he studied the growth of corn seedlings. In one group, he had 15 seedlings that were cross-pollinated, and in the other, he had 15 that were self-pollinated. After a fixed period, he recorded the final heights of each of the plants.

Cross	Self
23.5	17.4
12	20.4
21	20
22	20
...	...

Notice that these data *look* very similar to the correlations data that we saw in the previous section. They appear to be two columns of continuous data, but it's just the way they are arranged. We could instead have arranged them as a single column of continuous data with a column labelling the pollination method for each seedling, like this:

Seedling Number	Pollination Method	Height
1	Cross	23.5
2	Self	17.4
3	Cross	12
4	Cross	21
5	Self	20.4
...	...	...

To analyse these data you pool together the heights of all those plants that were cross-pollinated, work out the average height (the arithmetic mean) and compare this with the average height of those that were self-pollinated. You can display the results as a **Histogram**, like this:

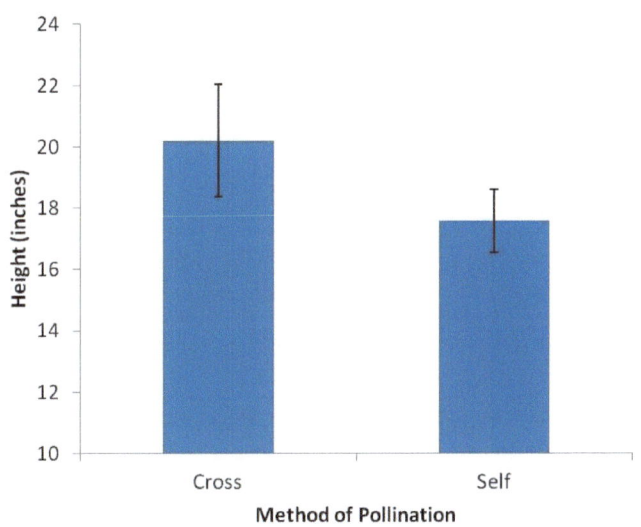

Histogram Showing the Mean Height of Darwin's Cross-Pollinated and Self-Pollinated Seedlings. Variation is shown by 95% Confidence Intervals.

Darwin found that the average height of the cross-pollinated plants was 20.19 inches, and 17.59 inches for the self-pollinated plants.

Of course, none of the plants were exactly 20.19 and 17.59 inches tall, there was a certain amount of variation in their heights. This variation can be expressed by the **standard deviation**, **variance** or other appropriate measure, like the **95% confidence intervals** shown.

Scientists tend to use standard deviations to show variation in measurements, but I prefer to use 95% confidence intervals because **95% confidence intervals are related to the statistical tests that you use** to decide whether there is likely to be a statistically significant difference between the measurements, so you can usually assess by eye whether there might be a difference.

Quite often, if there is a clear separation between the confidence intervals (in other words, little or no cross over between them) then there is likely to be a statistically significant difference between the groups of measurements.

By eye, there is a clear difference in the heights of the seedlings, and there is no cross over between the pairs of confidence intervals, so there is likely to be a significant difference between the heights of cross- and self-pollinated corn seedlings. We can't just check by eye, though, we also need to use a statistical test to tell us whether there is a statistically significant difference between the heights of the cross-pollinated and self-pollinated plants. Here, we can use either the **2-sample t-test** or the **Mann-Whitney U-test**.

The 2-sample t-test is used to test whether the <u>means</u> of two groups are the same, and assumes that the mean is the most appropriate **measure of centrality** for both groups of data. The 2-sample t-test is a more powerful test than the Mann-Whitney U-test as long as this assumption is not violated. The Mann-Whitney U-test, on the other hand, is used to test whether the <u>medians</u> of two groups are the same, and is used as an alternative to the 2-sample t-test when the mean is not the most appropriate measure of centrality for at least one of the groups of data.

So how do we know when the mean is appropriate to measure the centre point of the data?

For this, we need to run a **test of Normality**. There are a few of these, but we're going to use the **Anderson-Darling test of Normality**. If both groups of data pass this test, then we can safely quote the mean as being an appropriate measure of the central point of the data and use the 2-sample t-test to see if there is a statistically significant difference between the heights of the cross-pollinated and self-pollinated plants. If one or both groups of data fail this test, then we should quote the median and use the Mann-Whitney U-test to see if there is a significant difference in seedling heights.

Here is a **probability plot**, also known as a **Q-Q plot**, accompanied by the results of an Anderson-Darling test of Normality for the cross-pollination data:

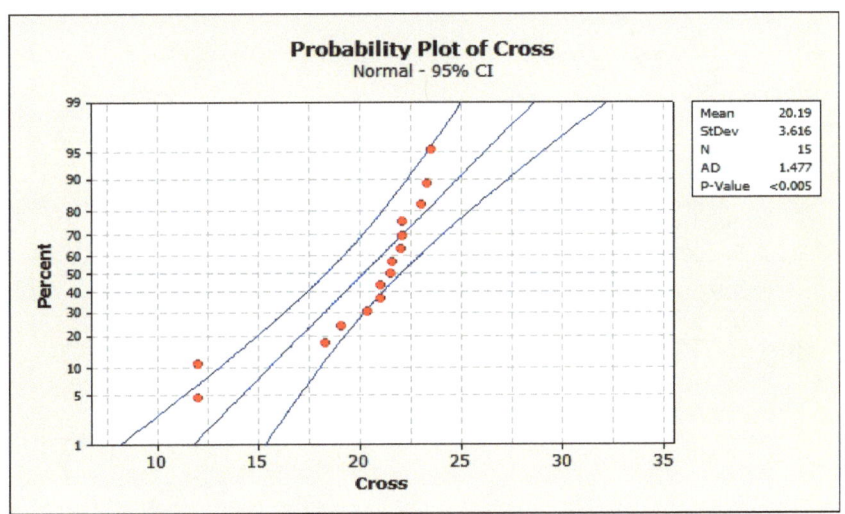

Probability Plot of Darwin's Cross-Pollination Corn Seedlings

The first thing to check is whether the plot points fall on a straight line (or nearly). They don't appear to, so these data may not be Normally distributed and we need to check the test p-value. If the p-value is large (>0.05) then the distribution is Normal and the mean is the appropriate measure of centrality. Here, p < 0.05, so we can be sure that the distribution is not Normal, we should not use the 2-sample t-test, but instead use the Mann-Whitney U-test.

The Mann-Whitney U-test orders all the observations from smallest to largest and then ranks them from 1 to N. In this way, the ranks can be compared across the two groups and the actual observations themselves are of no importance, so it doesn't matter that the data are non-Normal.

Here are the results of the Mann-Whitney U-test for the cross- and self-pollinated seedlings:

Mann-Whitney Test and CI: Cross, Self

```
         N   Median
Cross   15    21.5
Self    15    18.0
```

Point estimate for ETA1-ETA2 is 3.400
95.4 Percent CI for ETA1-ETA2 is (1.600, 4.901)
W = 305.5

Test of ETA1 = ETA2 vs ETA1 not = ETA2 is significant at 0.0026

The test is significant at 0.0026 (adjusted for ties)

The first thing to notice is that the medians are quoted rather than the means. Next, there are things called Eta 1 and Eta 2. These are the median values of each of the groups, and we are interested in the difference between them, which is 3.4. The 95% confidence interval of this difference is calculated, and what we're looking for here is whether these confidence intervals straddle zero. If they do, then there is not a significant difference between the medians. In this case, both the upper and lower 95% confidence interval bounds are positive, indicating that there is likely to be a significant difference in seedling heights.

Finally, we check the p-value to see if this is the case. As $p < 0.05$, we can conclude that Darwin's cross-pollinated corn seedlings were indeed significantly taller than the self-pollinated seedlings.

That's not quite the end of the story, though. Remember right at the beginning that we showed a histogram displaying the means and 95% confidence intervals? Well, we've just established that these are not appropriate measures to use, and we should instead use medians. When publishing our results, we should then use a **Box-and-Whiskers plot**, like the one overleaf.

In a Box-and-Whiskers plot, the box part is the **Inter-Quartile Range** (from the 1st quartile to the 3rd quartile of the data), and the bar across the middle is the median. The whiskers protruding out of the top and bottom of the box tell us the limits of our data. That's not quite the end of it though, because sometimes there are values that fall far outside the quartiles. These are the **outliers** and are usually plotted as separate points.

OK, so we've established that the data are not Normally distributed, so the medians are more appropriate than the means, and the Mann-Whitney U-test is more powerful than the 2-sample t-test for these data. We also discovered that Darwin's cross-pollinated corn seedlings were significantly taller than his self-pollinated seedlings.

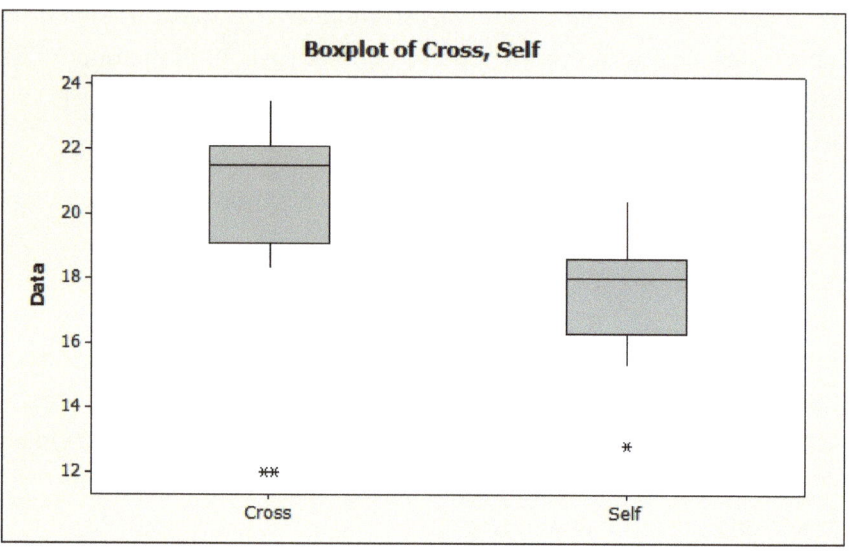

Box-and-Whiskers Plot of Darwin's Cross-Pollinated and Self-Pollinated Corn Seedlings

What if, on the other hand, we discovered that the data were Normally distributed? Well, then we can use the 2-sample t-test. Let's go ahead and do that, just to see where it takes us (overleaf).

Here, the means are quoted along with the standard deviation and the standard error of the mean. We're interested in the difference between the means, which is 2.61. The 95% confidence interval of this difference is calculated, and, as with the Mann-Whitney U-test, what we're looking for here is whether these confidence intervals straddle zero. Again, both confidence interval bounds are positive, and the p-value is significant, telling us that the cross-pollinated corn seedlings were significantly taller than the self-pollinated seedlings.

> **Two-Sample T-Test and CI: Cross, Self**
>
> Two-sample T for Cross vs Self
>
> | | N | Mean | StDev | SE Mean |
> |-------|----|-------|-------|---------|
> | Cross | 15 | 20.19 | 3.62 | 0.93 |
> | Self | 15 | 17.59 | 2.04 | 0.53 |
>
> Difference = mu (Cross) - mu (Self)
> Estimate for difference: 2.61
> 95% CI for difference: (0.38, 4.83)
>
> T-Test of difference = 0 (vs not =):
>   T-Value = 2.43
>   P-Value = 0.024
>   DF = 22

So, for these data it didn't matter whether we chose the most appropriate statistical test (the Mann-Whitney U-test) or the alternative (the 2-sample t-test). Both gave us the same result.

That doesn't mean that we can be complacent though. We must still endeavour to choose wisely, as this might not always be the case, and making a poor choice of statistical test could give us an incorrect result.

Test Yourself

- Look at the histogram and 95% confidence intervals above
    - What do you think they would look like if Darwin had performed this experiment with just 5 plants in each group rather than 15?
    - Would there have been a significant difference between the seedling heights?
    - What do you think would happen to the confidence intervals if Darwin had repeated this experiment with 30 plants in each group?
- Take another look at the Q-Q plot above.
    - If we were to remove the two outliers at the bottom left, do you think the remainder of the plot points fall on a straight line?
    - Would the 2-sample t-test then be more appropriate than the Mann-Whitney U-test?
    - Do you think it would be appropriate to have eliminated the outliers from the study?

Learn More

I don't want to come off as being some kind of a pest. I don't want you to think I'm *any* kind of a pest, but I bet you haven't visited the **Resources page** yet have you?

When you get to the end of this book, you'll read my bio, and when you do, you'll discover that **my mission is to unleash your inner data ninja**. It wouldn't be much of a mission if I couldn't get you to visit one little blog post where you can find all the resources you'll need to learn about associations and correlations, would it?

Don't forget, there's all sorts of FREE stuff there for you to find, so pop on over – I'll see you when you get there:

- http://chi2innovations.com/blog/discover-stats-blog-series/one-reason-your-correlation-results-are-probably-wrong/

Statistical Associations – Counts Within Categories

In the previous chapter, we looked at how to test for relationships when one of the variables was categorical; when we are dealing with the measurement of a variable within categories. Here we're going to learn how to test for relationships when both variables are categorical, that is, when we are dealing with counts in categories.

Let's go back and have another look at the data from the Body Fat *versus* Age study. These data were quite well behaved, but what if they weren't? Instead, let's pretend they were quite noisy, so much so that we didn't really trust the data, and we needed a way to cut through the noise. One way of doing this is to put the data into categories, like this:

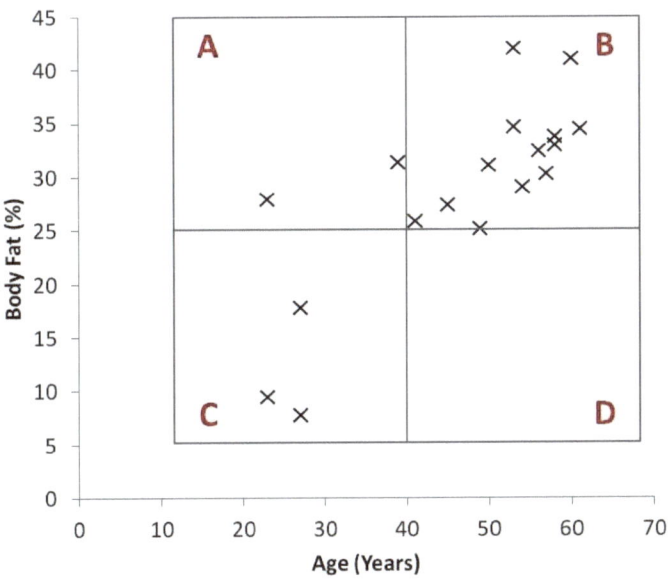

Scatter Plot of Age and Percentage of Body Fat with Four Distinct Quadrants

We could say that 40 years of age marks the boundary between Young and Old, and that 25% of body fat marks an important boundary between Hi and Lo. I've made these boundaries up, and really, they need to be based on experience and a knowledge of typical boundaries in your area of speciality.

What these boundaries do is separate your plot points into four distinct quadrants, as in the image on the previous page.

All you need to do now is add up all the points in each quadrant, and you'll get a **2x2 contingency table** (a contingency table is also known as a **cross-classification table**, **frequency table** or **confusion matrix**), like this:

		Age Years)	
		Young	Old
Body Fat (%)	Hi	2	13
	Lo	3	0

2x2 Contingency Table of Age and Body Fat

Of course, we don't *actually* form a box, place it over a scatter plot then add up all the points in each quadrant. That would be silly, particularly if we had hundreds of data points. Instead, we work with the data in the spreadsheet to transform our data into categories so they can be placed into their respective quadrants automatically. Then we collate the counts from each quadrant to make up the 2x2 table. When we've done this, our data might look something like this:

Age (Years)	AgeCat	Body Fat (%)	FatCat	Quadrant
23	Young	9.5	Lo	C
23	Young	27.9	Hi	A
27	Young	7.8	Lo	C
27	Young	17.8	Lo	C
39	Young	31.4	Hi	A
41	Old	25.9	Hi	B
...	...	...	...	...

OK, so let's return to our contingency table. Why have we arranged it to be a 2x2 table, when we could have had more categories for each variable? Good question!

Well, the 2x2 table is a special case of contingency tables, and the statistical tests we can use on the 2x2 table are much more powerful than those for larger tables. We'll look at that in a bit more detail soon.

What we're looking for in a 2x2 table is whether the large numbers fall on a diagonal. If they do, that means that there is likely to be some sort of trend in the data (remember the regression line – we're looking for a best-fit line that is not horizontal or vertical). For our data, the largest two numbers are on the diagonal from bottom-left to top-right. It looks like there might be a trend that indicates that older people are more likely to have higher levels of body fat.

How do we know if this is true or not? We need to perform a statistical hypothesis test. I've run a couple of different tests below so that we can compare them and their appropriateness for these data:

	P-value		
	Left tail (negative)	Right tail (positive)	2-tail (both)
Chi-Squared Test			0.002
Fisher's Exact Test	1.000	0.012	0.012

Selection of Statistical Results for Age *versus* Body Fat

Typically, the **Chi-Squared Test** is used for the analysis of contingency tables. Actually, the Chi-Squared Test can be used for tables of any size and is simple to compute, even with an Excel spreadsheet or even a pen and paper, although I'm not going to explain how to do it here.

The result of the Chi-Squared Test (p = 0.002) tells us that there is a statistically significant relationship between our variables.

There are a couple of problems with this calculation though. Firstly, the Chi-Squared Test is less accurate when there are counts of less than five in any of the cells. Worse than that, it really shouldn't be used at all when there are counts of zero in any cell. Oops, our data seem to have violated all the assumptions of the Chi-Squared Test. It is fair to say that we can't trust the result!

As an alternative to the Chi-Squared Test, we can use the **Fisher's Exact Test**. This is computationally more complex than the Chi-Squared Test (for manual computation – it is not an issue for modern computers), but it is usually more accurate, is well suited to small sample sizes, and has no issues with small cell counts – even zeros. The downside is that it can only be used for 2x2 contingency tables.

The result of the Fisher's Exact Test (p = 0.012) tells us that there is a statistically significant relationships between our variables. No assumptions are violated and we can trust the result. Hooray!

But hold on a minute – why are there three different p-values for the Fisher's Exact Test?

When you have no prior hypothesis about what the outcome might be, then you use the 2-tail result. On the other hand, if you have a prior hypothesis, as we did with the Age v Body Fat experiment, then you state your hypothesis up front and then use the appropriate 1-tail result. Note that if your hypothesis was that there would be a positive association between your variables, then you MUST use the right-tail result. No cheating! But what if the right-tail result is not significant and the left-tail result is? Tough luck – your hypothesis was the wrong way round!

In our case we hypothesised that as you get older you will have more body fat – a positive association – then we must use the right-tail result, which is significant. Phew!

The good thing here is that the 1-tail result usually has a much smaller p-value than the 2-tail, so having a prior hypothesis really helps you pin down the result you're looking for.

For our data, the right-tail and 2-tail results are the same, so it didn't matter whether we had a prior hypothesis or not – the result was that as you age you are likely to gain extra body fat, and this result is statistically significant.

Remember that we started out with continuous data for Age and Body Fat, and we discovered that there was a significant correlation. On transforming these data to categories and analysing by contingency table, we found the same result, that there is a significant association.

So if we can get the same result by either method, which one should we use? Well, it all depends on your data. Continuous data contains more information than categorical data, but it also contains more bias, variation and noise. Transforming your data from continuous to categorical gets rid of a lot of these problems, but you also lose a lot of the information too.

What I'm getting at here is that analysing data in a continuous form is preferable to transforming it to categories, but only when the data is well behaved enough so that the 'true' answer shines through. How do we know when this is the case? Experience!

Test Yourself

- Take another look at the scatter plot with the four quadrants.
    - If we had categorised Age, but not Body Fat, what type of analyses would you use with these data?
    - What types of graph would be appropriate?
    - Which statistical tests would be the right ones to use?
- Visualise what the outcome would be.
    - Would there have been a significant association with these data?
    - What would be the approximate measurements of the descriptive statistics?

Learn More

OK, so you know the drill by now – this is the bit where I remind you to visit the **Resources page**, so here it is. Visit the resources page:

- http://chi2innovations.com/blog/discover-stats-blog-series/one-reason-your-correlation-results-are-probably-wrong/

Your Correlation Results Are Probably Wrong!

What we've learnt so far is that whether you have two continuous variables, two categorical variables or one of each, finding a statistical relationship between the pair of them is pretty straightforward. It's easy to see that where you have maybe a dozen variables you can analyse them pairwise using the correlation and association tests from above and find out which relationships exist in your dataset, and which do not.

At this point, you might be starting to feel comfortable, a little confident, and perhaps even a tiny bit smug.

Not so fast, cowboy – it's not quite that simple!

You see, when you analyse a pair of variables using the **univariate tests** that we discussed above, you're testing to see whether there is a relationship between these variables *without taking into account any other potential factors*. There are loads of ways in which your variables might be interacting with and influencing each other, so when you have a significant p-value from univariate analysis you can't be sure that the answer you get is correct.

> **UNIVARIATE STATS**
>
> Univariate statistics are tests that you use when you are comparing variables one at a time with your hypothesis variable.
>
> In other words, you compare this with that whilst ignoring all other potential influences.

Let me make it easy for you. If you get a non-significant p-value (larger than 0.05), you can be pretty sure (actually, 95% sure) that there is not a direct relationship between your variables. That's not to say that one does not *influence* the other indirectly, it may do, but there is not likely to be an independent relationship between them. On the other hand, if you get a significant p-value (smaller than 0.05), the best you can say is that there *may* be a relationship between them. The relationship *might* be independent, but equally it might not. This flow chart might help you a little:

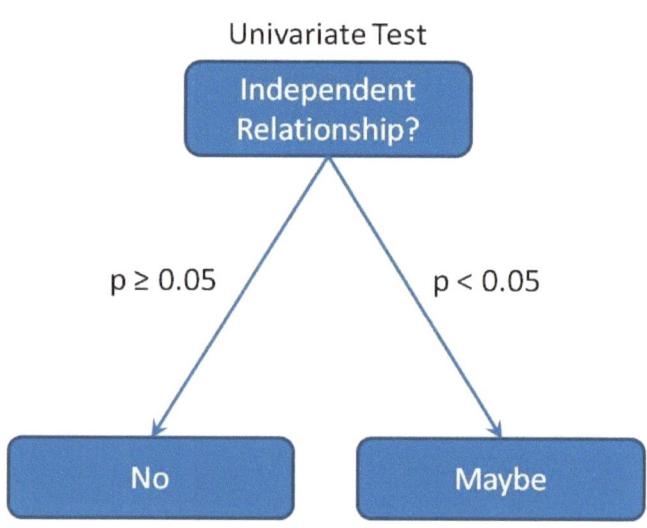

Feeling a little less smug now, aren't we?

So if univariate tests don't give us the answers we need, where do we go from here? Well, the univariate tests are still useful to us. Remember that they're pretty good at telling us when there *isn't* a direct relationship between a pair of variables. This is useful information and allows us to narrow the field of possibilities between what *might* be related to your main variable (aka **hypothesis variable**) and which ones aren't.

So, in turn you test each variable against your hypothesis variable to see which of them are *not* related. Then you discard them. What remains are the variables that might be related to it.

The next step gets tricky because we now need to test the relationship between the hypothesis variable and all of these variables whilst taking into account all the possible interactions between them. Sounds scary!

We're now dipping our toes into the world of **multivariate analysis**.

> **MULTIVARIATE STATS**
>
> You use multivariate statistics when you want to measure the relative influence of many variables on your hypothesis variable.
>
> In other words, when you simultaneously compare *this*, *that* and *the other* against your target.

So what does a multivariate test look like? Well, here's an example, where I've regressed Age and Weight against a hypothesis variable Systolic Blood Pressure:

> **Regression Analysis: Systolic Blood Pressure versus Age, Weight**
>
> The regression equation is
> Systolic Blood Pressure = 30.990 + (0.861 x Age) + (0.335 x Weight)
>
> | Predictor | Coef | SE Coef | T | P |
> |-----------|--------|---------|------|-------|
> | Constant | 30.990 | 11.94 | 2.59 | 0.032 |
> | Age | 0.861 | 0.248 | 3.47 | 0.008 |
> | Weight | 0.335 | 0.131 | 2.56 | 0.034 |
>
> S = 2.318   R-Sq = 97.7%

If you remember back to when we did a univariate regression test, the format of the result is the same. What is different here is that there are two variables that are significantly related to Systolic Blood Pressure. The result here tells us that the variables Age and Weight are independent of each other and they are both significant factors in high Systolic Blood Pressure – the older and heavier you are, the higher your Systolic Blood Pressure is likely to be.

It is important that you learn to use both univariate and multivariate tests because although univariate stats are easier to use and they give you a good *feel* for your data, they don't take into account the influence of other variables so you only get a partial – and probably misleading – picture of the story of your data.

On the other hand, multivariate stats *do* take into account the relative influence of other variables, but these tests are much harder to understand and you don't get a good feel for your data.

My advice is to **do univariate analyses on your data first to get a good understanding** of the underlying patterns of your data, then **confirm or deny these patterns with the more powerful multivariate analyses**. This way you get the best of both worlds and when you discover a new relationship, you can have confidence in it because it has been discovered and confirmed by two different statistical analyses.

When pressed for time I've often just jumped straight into the multivariate analysis. Whenever I've done this, it has always ended up costing me more time – I find that some of the results don't make sense and I have to go back to the beginning and do the univariate analyses before repeating the multivariate analyses.

I advise that you think like the tortoise rather than the hare – slow and methodical wins the race...

Test Yourself

- Take another look at the multivariate regression result from above.
    - What would be the effect on Systolic Blood Pressure if one or both of the coefficients of Age and Weight were negative?
    - What should you do if one of the variables has a p-value greater than 0.05?
    - If a third variable was also found to be significant, how would that change the regression equation?

Learn More

Now we're getting near to the end of the book. So far, at the end of every chapter I've put in a reminder for you to visit the **Resources page**. I'm not going to put a reminder at the end of this chapter. I'm not. You've got the message. No more reminders to **visit the Resources page**. Not even any subliminal ones.

I'm not going to remind you of the FREE book **Correlation Is Not Causation**. Nor of all the other FREE stuff that you can get. And I'm not even going to think about the video courses.

Finally, I'm not going to remind you that this is the link to the resources page:

- http://chi2innovations.com/blog/discover-stats-blog-series/one-reason-your-correlation-results-are-probably-wrong/

Bonus: Automating Associations and Correlations

If you've ever had to analyse a dataset to find the associations and correlations you'll know how time consuming it can be. A quick 'let's have a little look and see what we can find' in a small dataset with a couple of dozen variables (columns) can often take a few months of analysis, and a full investigation can take up to a year. Larger datasets just can't be analysed adequately with the current crop of commercial stats programs available.

Fortunately, there is a solution.

CorrelViz is a fully automated program that allows you to visualise all of your correlations.

It cleans and classifies your data, screens potential relationships by using the appropriate univariate statistical tests, confirms or denies these relationships with multivariate stats, and presents you with an intuitive, interactive visualisation of the story of your data.

All this is completely automatic and takes you **from data to story in minutes, not months**.

Best of all, CorrelViz flips the methodology of data analysis on its head. Instead of doing your analysis first and then trying to visualise your results last, in CorrelViz, the first thing you see is an interactive visualisation of all the relationships in the dataset. **You get the story *first*.** All of it!

Through all this, there is no manipulation of data, no selecting incorrect statistical tests, no worries about confounding variables, and the whole process takes just minutes. Extract the relevant results and your whole research effort could be over before the skin on your skinny latte has cooled off!

Best of all – you can use it for **FREE**.

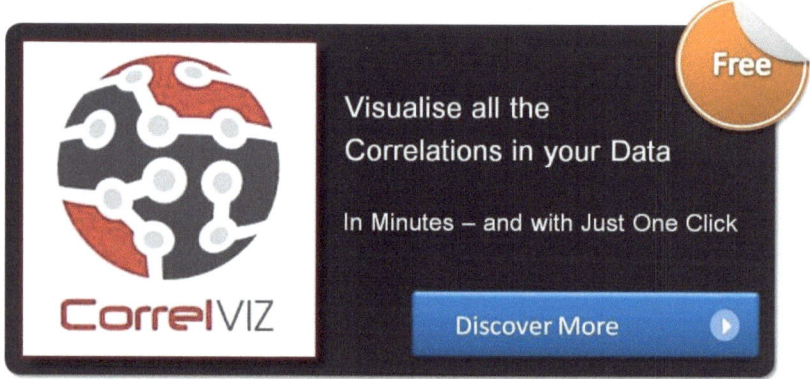

You can find CorrelViz here:

- http://chi2innovations.com/correlviz/

Learn More

OK, so you've managed to make it to the end. You're wondering what comes next.

'Where can I learn more about associations and correlations?' I hear you asking yourself.

Well, I'm guessing that you know what your next step is. I've put some very subtle clues throughout the book, and you might have missed them, but your next step is to **visit the Resources page**.

This is the final reminder, honest.

Over at the **Resources page** you'll be able to get all sorts of great learning materials to help you take your next step in learning about **associations**, **correlations**, **regressions**, **univariate statistics**, **multivariate statistics** and lots more. There are **books**, **blogs**, **video courses** and all sorts of stuff, and they're updated regularly so you'll always find the best of the web right there.

Here is the link to the resources page:

- http://chi2innovations.com/blog/discover-stats-blog-series/one-reason-your-correlation-results-are-probably-wrong/

Summary

So now, we really have reached the end of the **Beginner's Guide to Correlation Analysis**, and I hope you got something useful out of it.

We discovered that:

- associations and correlations are really just statistical tools to find out whether there is evidence of the existence of a relationship between variables
- statistical significance can often be judged by eye if an appropriate visualisation is used (but you still need to run the test to be sure!)
- having a holistic strategy of univariate and multivariate tests to discover all the relationships in your dataset will make your analysis go that bit smoother

I hope that you'll check out my other books in the **Bite-Size Stats Series**. You can either find them in the Resources page, at Amazon or in many other online bookstores.

If you enjoyed **Beginner's Guide to Correlation Analysis** and you want to know when the next one will be out, just visit the Resources page. You'll see a button at the top of the page where you can **subscribe to our Newsletter** – I promise you'll be the first to know!

Here is the link to the resources page:

- http://chi2innovations.com/blog/discover-stats-blog-series/one-reason-your-correlation-results-are-probably-wrong/

###

About The Author

Lee Baker is an award-winning software creator that lives behind a keyboard in a darkened room. Illuminated only by the light from his monitor, he aspires to finding the light switch.

With decades of experience in science, statistics and artificial intelligence, he has a passion for telling stories with data. Despite explaining it a dozen times, his mother still doesn't understand what he does for a living.

Insisting that data analysis is much simpler than we think it is, he authors friendly, easy-to-understand books that teach the fundamentals of data analysis and statistics.

His mission is to unleash your inner data ninja!

As the CEO of Chi-Squared Innovations, one day he'd like to retire to do something simpler, like crocodile wrestling.

Claim Your FREE eBook Now!

Correlation Is Not Causation is the sister book to **Beginner's Guide to Correlation Analysis**, and shows you how to avoid the 5 correlation-causation traps that even the pros fall into.

Download your **FREE** copy right here:

http://bit.ly/2FetzGT

Leave a Review

Thank you for reading **Beginner's Guide to Correlation Analysis**.

I hope you enjoyed reading it as much as I enjoyed writing it. If you did, please take a moment to return to where you purchased this book and leave a review.

Thank you!

Lee Baker

www.ingramcontent.com/pod-product-compliance
Lightning Source LLC
Chambersburg PA
CBHW040333220526
45473CB00009B/2669